I0480735

This Book Belongs To:

_____

Copyright © 2020
All rights reserved. This book or any portion thereof
may not be reproduced or used in any manner whatsoever
without the express written permission of the publisher.

Color Test Page

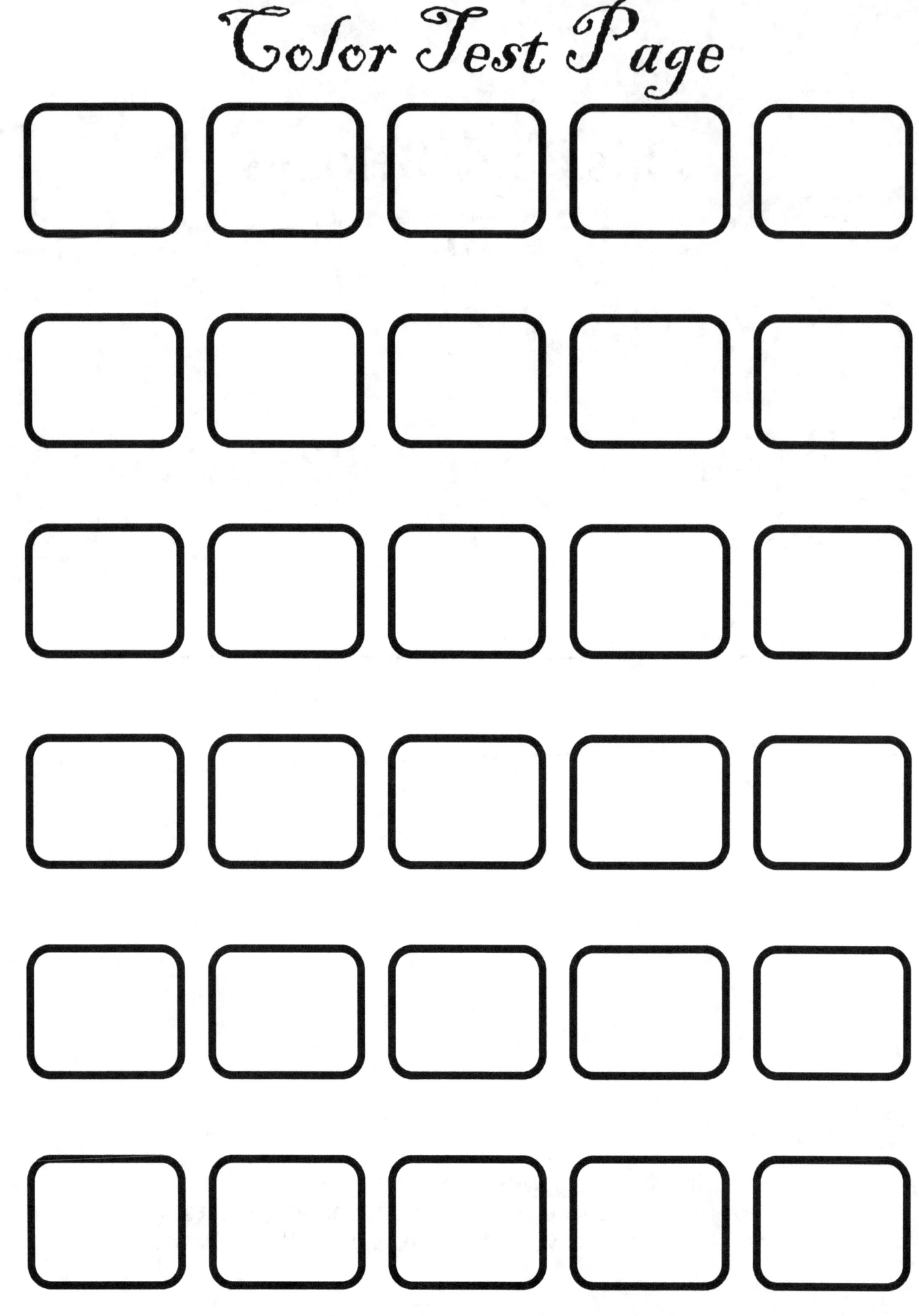

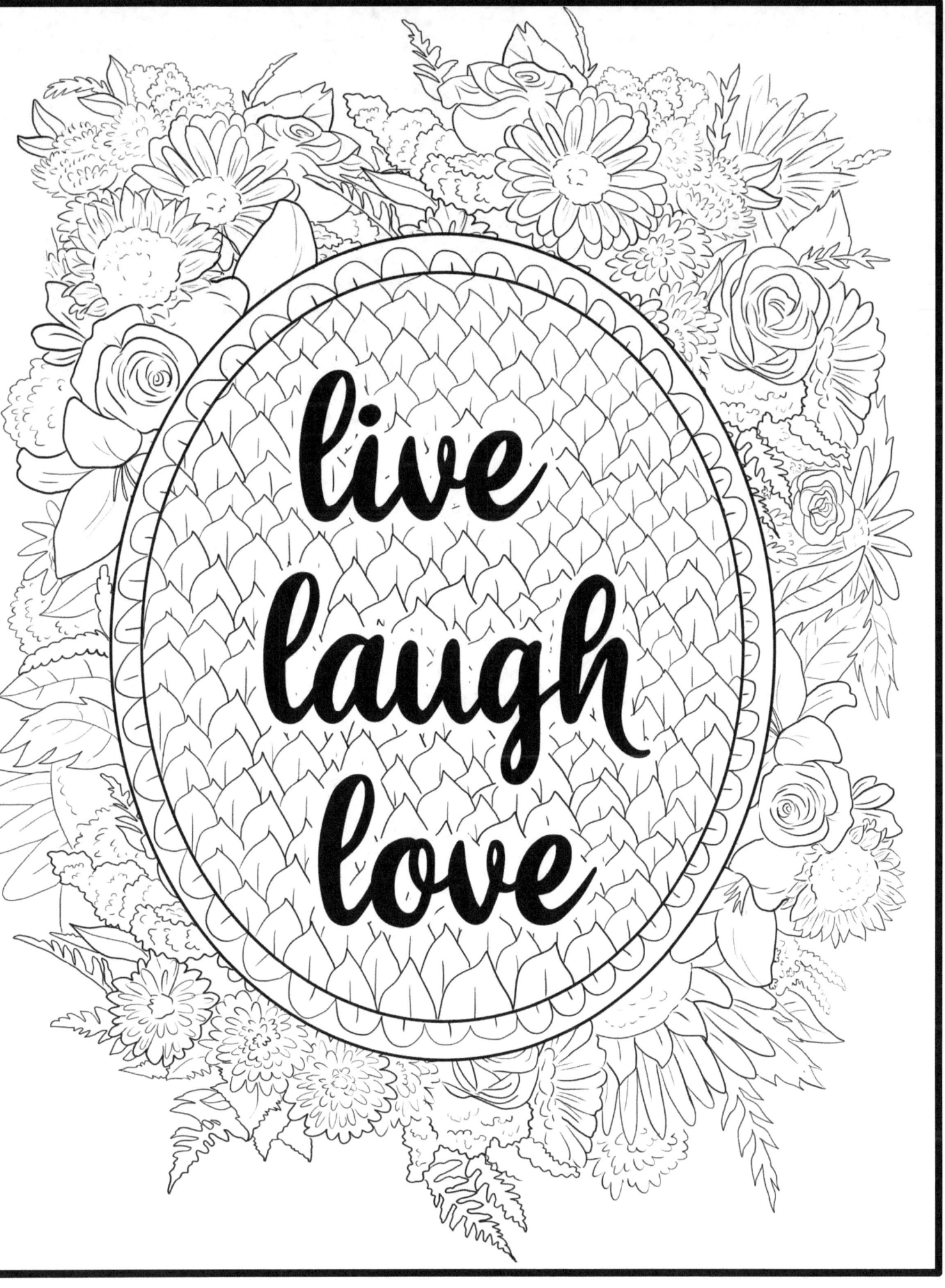

As the sun shines both on the cedar tree and the smallest flower, so the Divine sun illumines each soul.

Therese of Lisieux

www.ingramcontent.com/pod-product-compliance
Lightning Source LLC
Chambersburg PA
CBHW081542220526
45467CB00010B/3291

www.ingramcontent.com/pod-product-compliance
Lightning Source LLC
Chambersburg PA
CBHW081542220526
45467CB00010B/3291